easy learning

Mental maths

Ages 7–9

8 + 3 + 9 = ☐

76 − 4 = ☐

92 × 100 = ☐

81 ÷ 9 = ☐

Peter Clarke

How to use this book

- Find a quiet, comfortable place to work, away from distractions.
- Ask your child what maths topic they are doing at school and choose an appropriate topic.
- Tackle one topic at a time.
- Help with reading the instructions where necessary and ensure that your child understands what they are required to do.
- Help and encourage your child to check their own answers as they complete each activity.
- Discuss with your child what they have learnt.
- Let your child return to their favourite pages once they have been completed, to play the games and talk about the activities.
- Reward your child with plenty of praise and encouragement.

Special features

- Yellow boxes: Introduce and outline the key maths ideas.
- Example boxes: Show how to do the activity.
- Orange shaded boxes: Offer advice to parents on how to consolidate your child's understanding.
- Games: Some of the topics include a game, which reinforces the topic. Some of these games require a spinner. This is easily made using a pencil, a paperclip and the circle printed on each games page. Place the pencil and paperclip at the centre of the circle and then flick the paperclip to see where it lands.

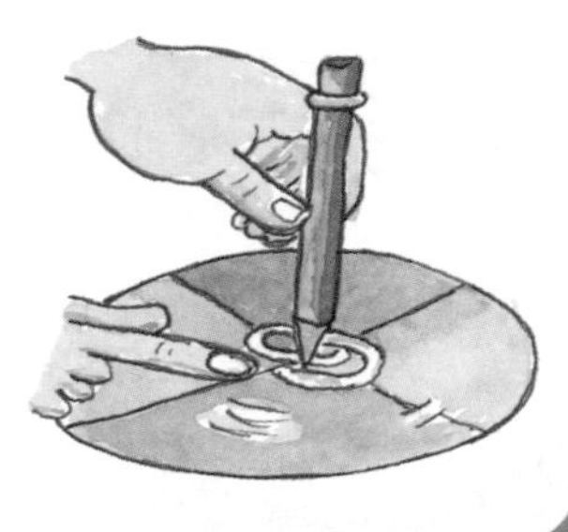

Published by Collins
An imprint of HarperCollins*Publishers*
1 London Bridge Street
London SE1 9GF

Browse the complete Collins catalogue at www.collins.co.uk

10 9 8 7 6

ISBN 978-0-00-813423-5

The author asserts his moral right to be identified as the author of this work.

The author wishes to thank Brian Molyneaux for his valuable contribution to this publication.

British Library Cataloguing in Publication Data

A Catalogue record for this publication is available from the British Library

Written by Peter Clarke
Page design by G Brasnett, Cambridge and Contentra Technologies
Illustrated by Kathy Baxendale, Rachel Annie Bridgen and Graham Smith
Cover design by Sarah Duxbury and Paul Oates
Cover illustration © Leo Blanchette/Shutterstock.com
Project managed by Chantal Peacock and Sonia Dawkins

Contents

Numbers 1

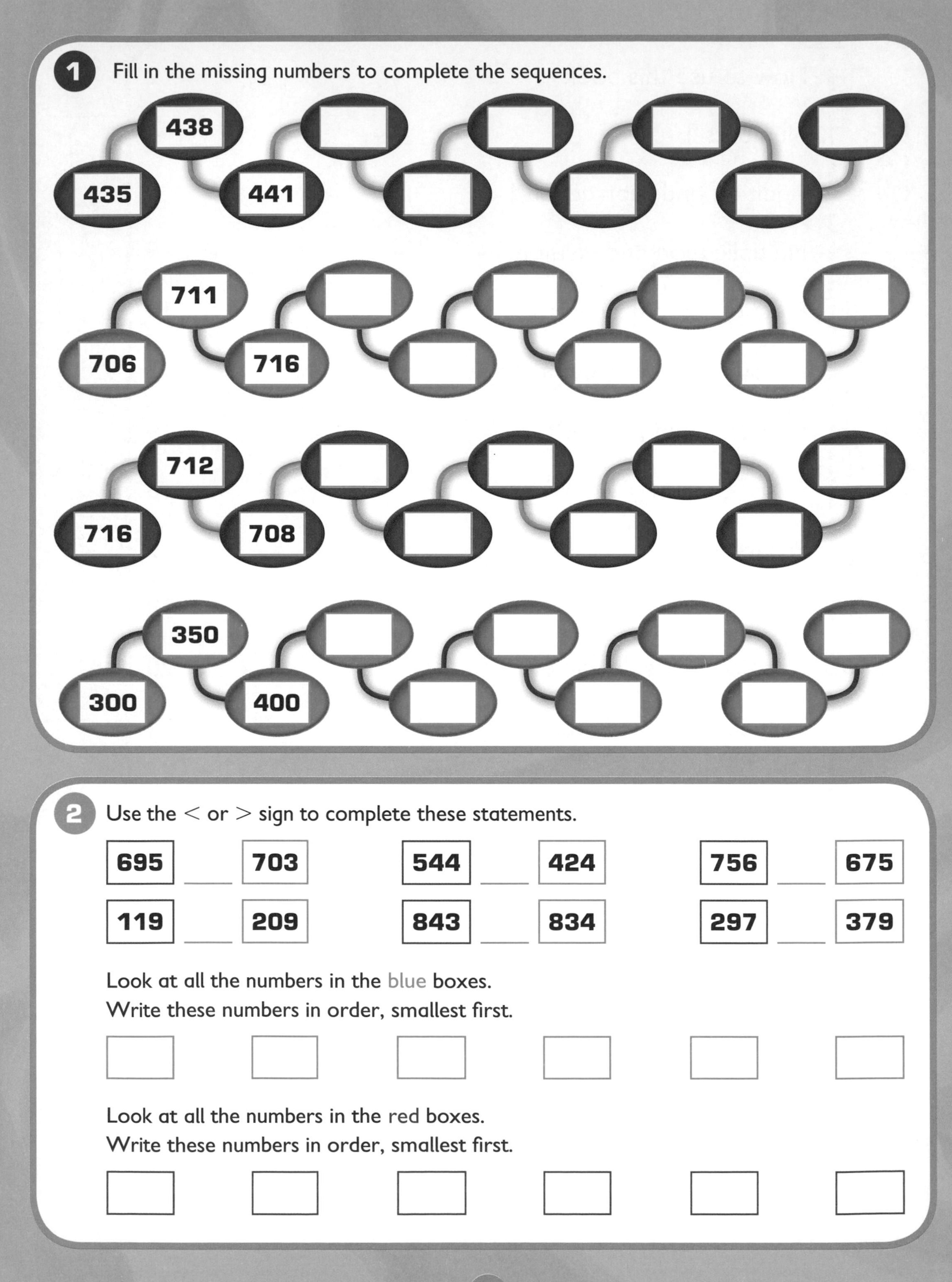
1 Fill in the missing numbers to complete the sequences.
435
438
441
706
711
716
716
712
708
300
350
400
2 Use the < or > sign to complete these statements.
695
703
544
424
756
675
119
209
843
834
297
379
Look at all the numbers in the blue boxes.
Write these numbers in order, smallest first.
Look at all the numbers in the red boxes.
Write these numbers in order, smallest first.

Game: Comparing numbers

You need: pack of playing cards with the Jacks, Queens and Kings removed, counters

- Shuffle the cards and place them face down in a pile.
- Take turns to:
  - pick the top three cards
  - arrange the cards to make a three-digit number and read it out, e.g.
  - place one of your counters on a box in the grid that describes your number.

- If you can't find a description in the gird that matches, miss that turn.
- After each player has had a turn, collect up all the cards and shuffle them again.
- The winner is the first player to complete a line of four counters. A line can go horizontally or vertically.

>700	less than 850	>100	>200	<400
less than 800	<650	>400	>300	more than 600
less than 750	<200	more than 250	more than 650	less than 350
<600	>550	less than 700	>800	<250
less than 450	more than 500	<300	<500	more than 800

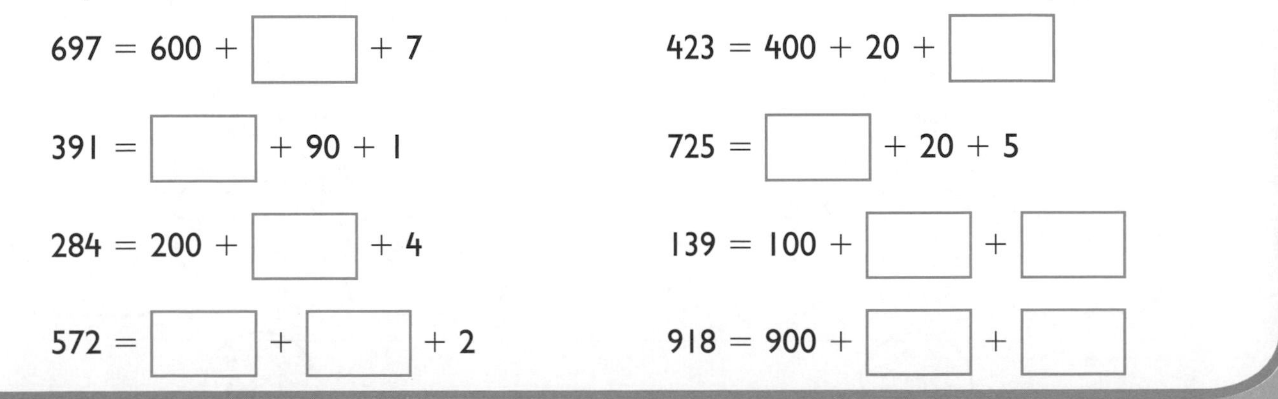

3 Complete each number sentence.

697 = 600 + ☐ + 7

423 = 400 + 20 + ☐

391 = ☐ + 90 + 1

725 = ☐ + 20 + 5

284 = 200 + ☐ + 4

139 = 100 + ☐ + ☐

572 = ☐ + ☐ + 2

918 = 900 + ☐ + ☐

In order to calculate with numbers, it is important that your child is able to count, recognise, read, write, compare, order and round numbers to 1000 then 10 000. They also need to have a secure understanding of place value, i.e. 582 = 500 + 80 + 2.

Addition and subtraction 1

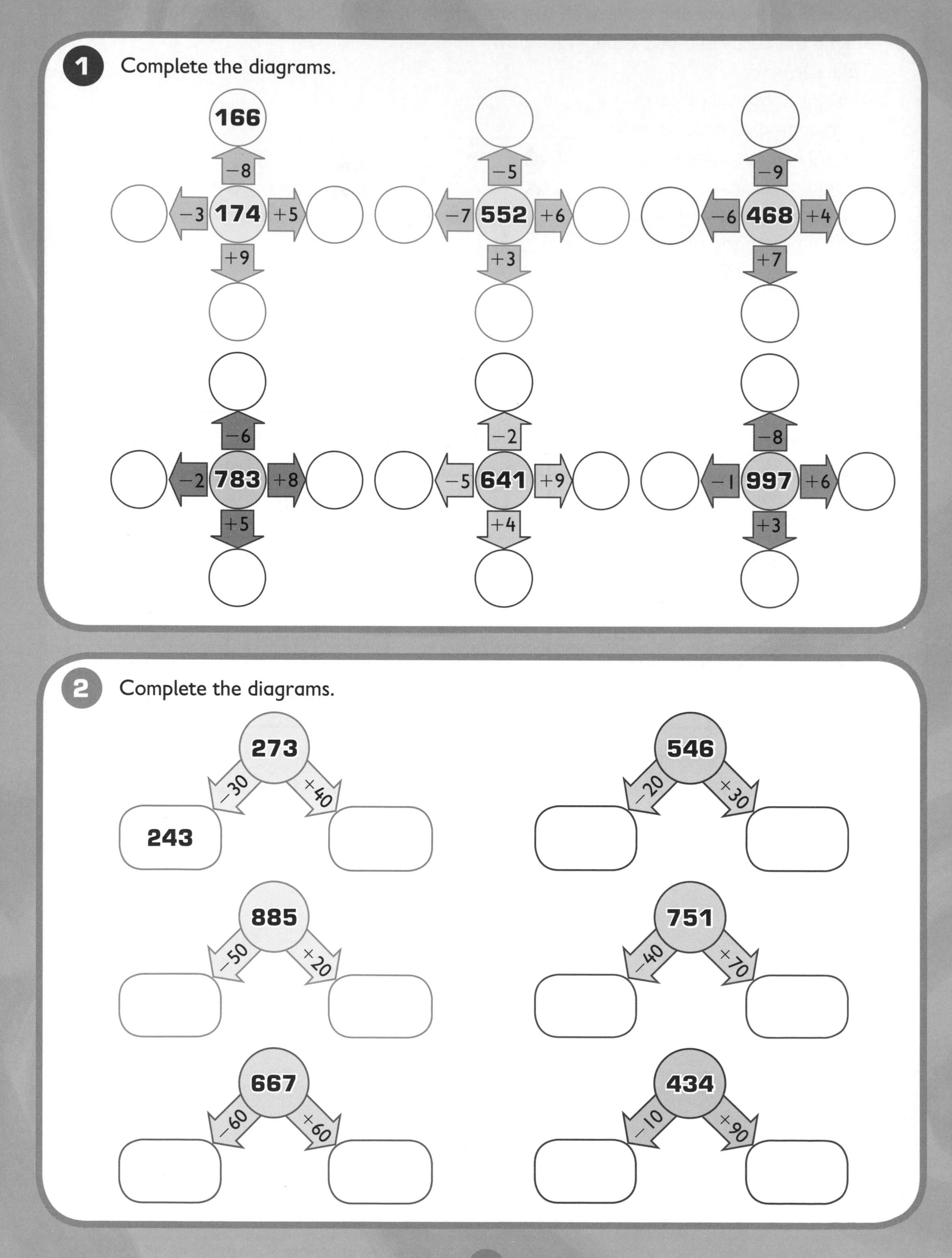

Game: Spin the winner

You need: 2 pencils and 2 paperclips (for the spinners)

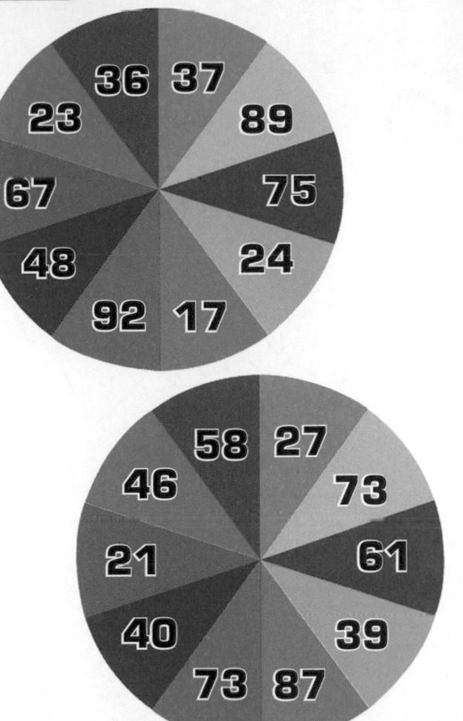

Game 1: Add the numbers

- Take turns to spin both spinners and add the two numbers together.
- The player with the larger total wins that round.
- The overall winner is the first player to win six rounds.

Game 2: Find the difference

- Take turns to spin both spinners and find the difference between the two numbers.
- The player with the smaller difference wins that round.
- The overall winner is the first player to win six rounds.

3 Complete the diagrams.

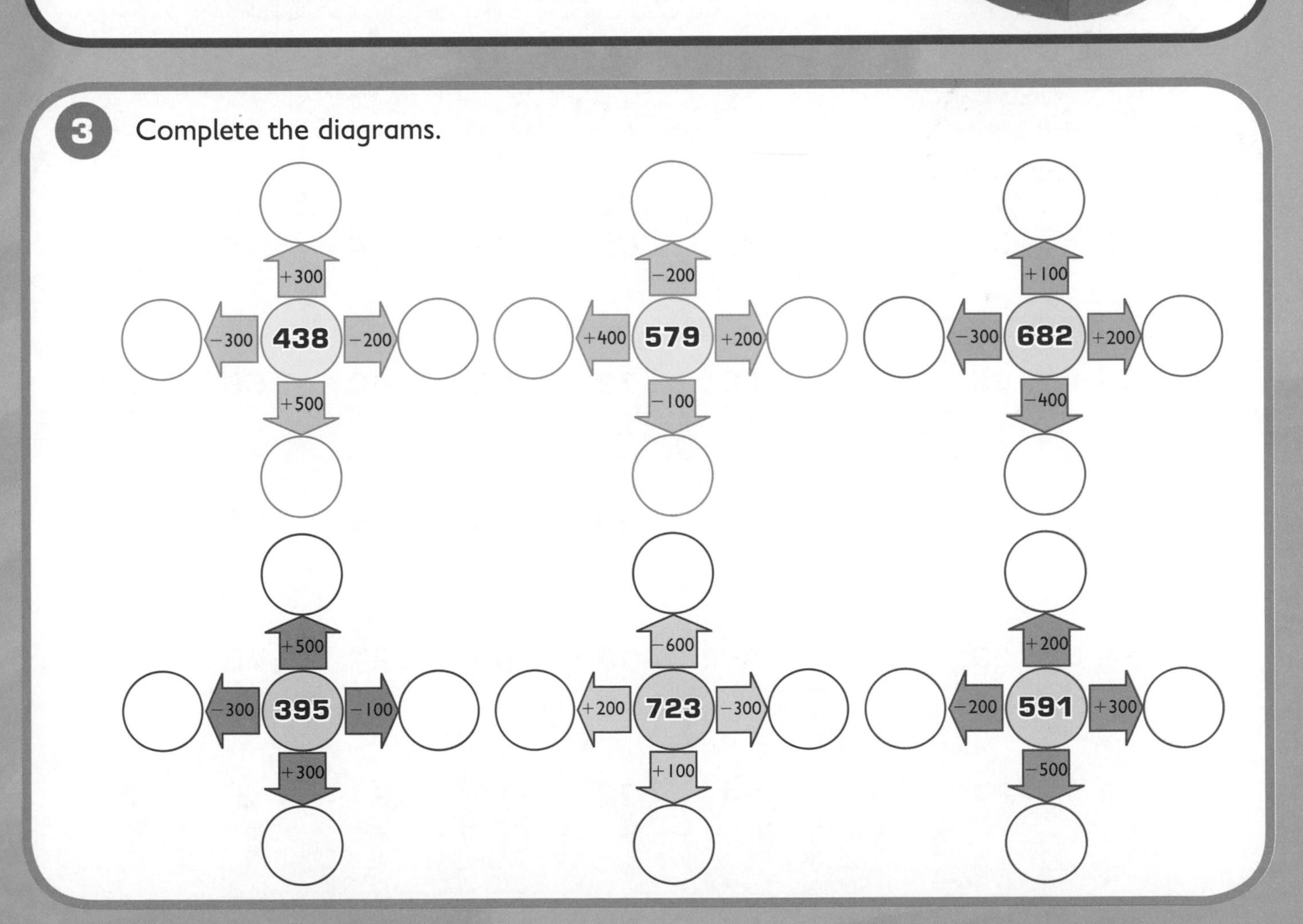

Being able to confidently add and subtract a one-digit number and a multiple of 10, to or from a two- or three-digit number will help your child when adding and subtracting larger numbers such as 268 + 47 and 573 − 25.

Multiplication and division 1

1 Complete the multiplication rings.

×3: 7, 10, 6, 9, 3, 4, 8, 5 (21)

×12: 2, 9, 5, 8, 3, 4, 10, 7

×6: 2, 8, 5, 3, 4, 9, 6, 7

×7: 8, 3, 10, 6, 2, 4, 7, 5

×9: 7, 5, 10, 3, 8, 4, 6, 9

×8: 6, 10, 8, 2, 9, 4, 7, 5

2 Complete each bubble puzzle.

÷6	42	36	54	30	66	60	24	48	18
	7								
÷8	24	80	48	96	72	64	40	56	32
÷4	20	8	36	28	40	16	24	32	48
÷9	36	99	81	45	63	90	27	72	54
÷3	12	27	18	24	33	15	21	9	30
÷7	63	70	84	28	56	49	35	42	21

Game: Quick tables

You need: 2 pencils and 2 paperclips (for the spinners), 20 counters

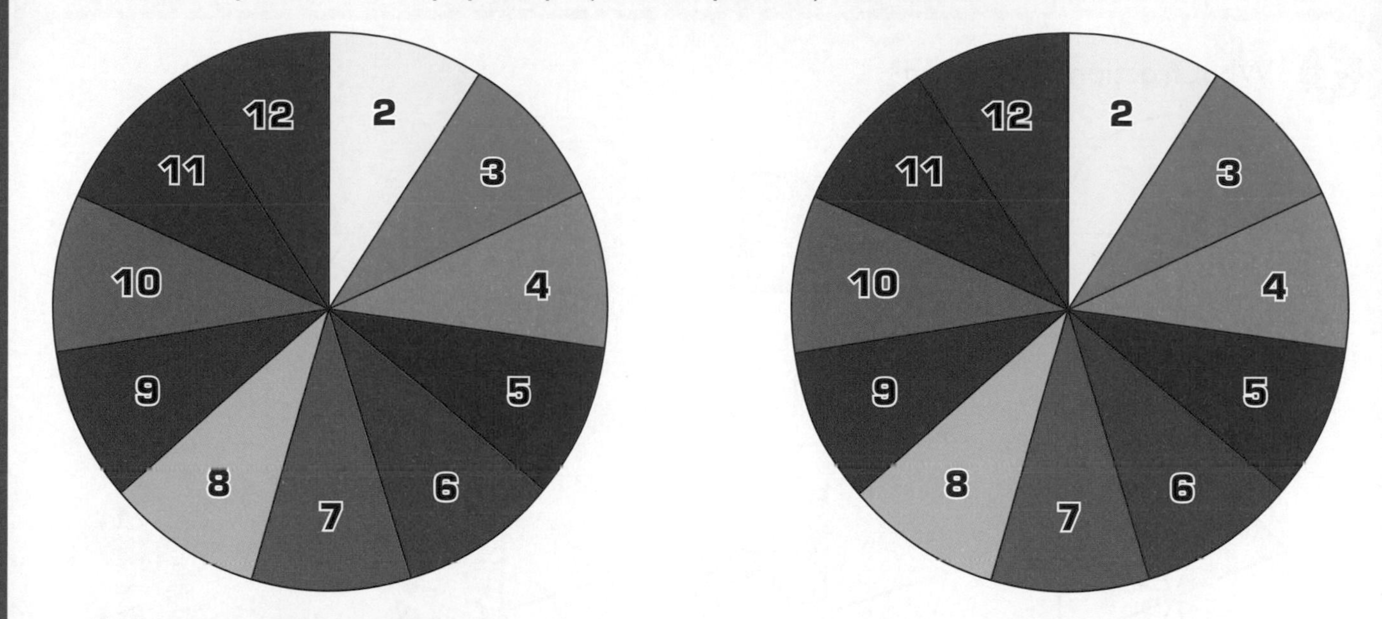

- Each player spins a spinner.
- Both players multiply the two numbers together.
- The first player to call out the correct answer wins that round and takes a counter.
- The overall winner is the first player to collect 10 counters.

3 Complete each tower.

×10		×100
670	**67**	
	52	
	83	
	24	
	75	
	38	
	91	

÷10		÷100
30	**300**	
	8700	
	5200	
	6000	
	1800	
	2900	
	400	

Being able to recall the answers to the times tables facts up to 10 × 10, and the related division facts, will help your child when multiplying and dividing larger numbers such as 34 × 6 and 83 ÷ 4.

Fractions and decimals 1

1 What fraction is coloured?

2 Use the < and > signs to compare these fractions.

$\frac{1}{2}$ ___ $\frac{1}{3}$ $\frac{1}{6}$ ___ $\frac{1}{4}$ $\frac{1}{5}$ ___ $\frac{1}{8}$

$\frac{1}{10}$ ___ $\frac{1}{6}$ $\frac{1}{9}$ ___ $\frac{1}{7}$ $\frac{1}{3}$ ___ $\frac{1}{5}$

Look at the fractions in the blue boxes. Write the fractions in order, smallest first.

Look at the fractions in the red boxes. Write the fractions in order, smallest first.

3 Add or subtract each pair of fractions.

$\frac{5}{7} + \frac{1}{7} =$ $\frac{3}{4} + \frac{1}{4} =$ $\frac{7}{11} - \frac{3}{11} =$

$\frac{7}{9} - \frac{2}{9} =$ $\frac{6}{7} - \frac{4}{7} =$ $\frac{1}{9} + \frac{4}{9} =$

Game: Finding fifths

You need: 2 pencils and 2 paperclips (for the spinners), counters

- Take turns to:
  - spin both spinners
  - multiply the number by the fraction
  - put a counter on the answer on the grid.
- If you can't find the answer on the grid, miss that turn.
- The winner is the first player to complete a line of three counters. A line can go horizontally, vertically or diagonally.

40	9	12	4	8	16	5
10	15	20	6	18	24	7
14	21	28	32	27	12	3
30	11	22	33	44	36	48

4 Work out the cost of the following.

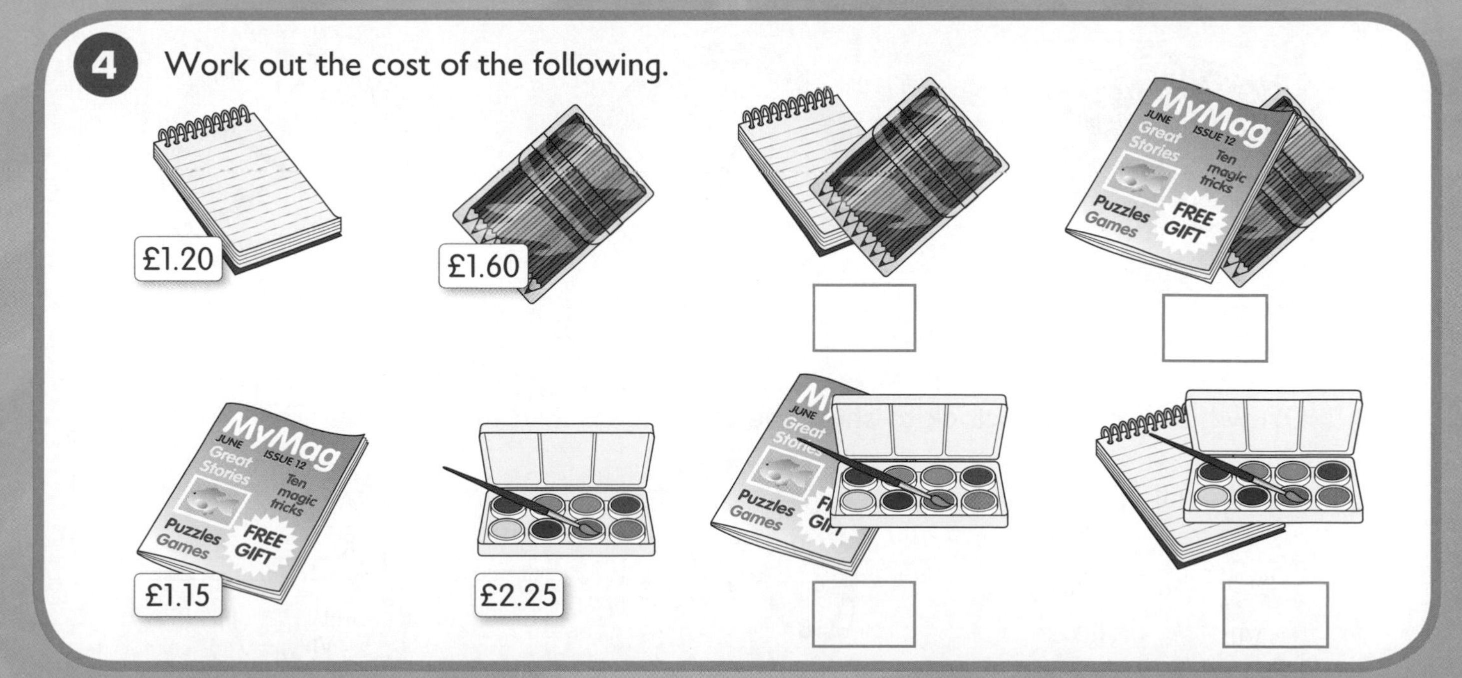

Look for examples of fractions and decimals in newspapers or magazines, around the home, or at the shops. Ask your child to explain to you what they mean.

Measurement and geometry 1

1 Complete the table.

1 2 3 4 5 6 7

Shape	Name of shape	Number of sides	Number of vertices	Number of lines of symmetry
1				
2				
3				
4				
5				
6				
7				

2 What time does each clock show?

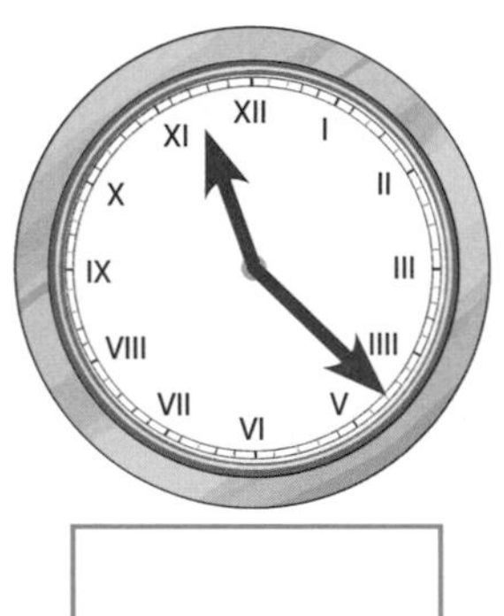

Draw hands on each clock to show the time.

2:28 **5:52** **10:41**

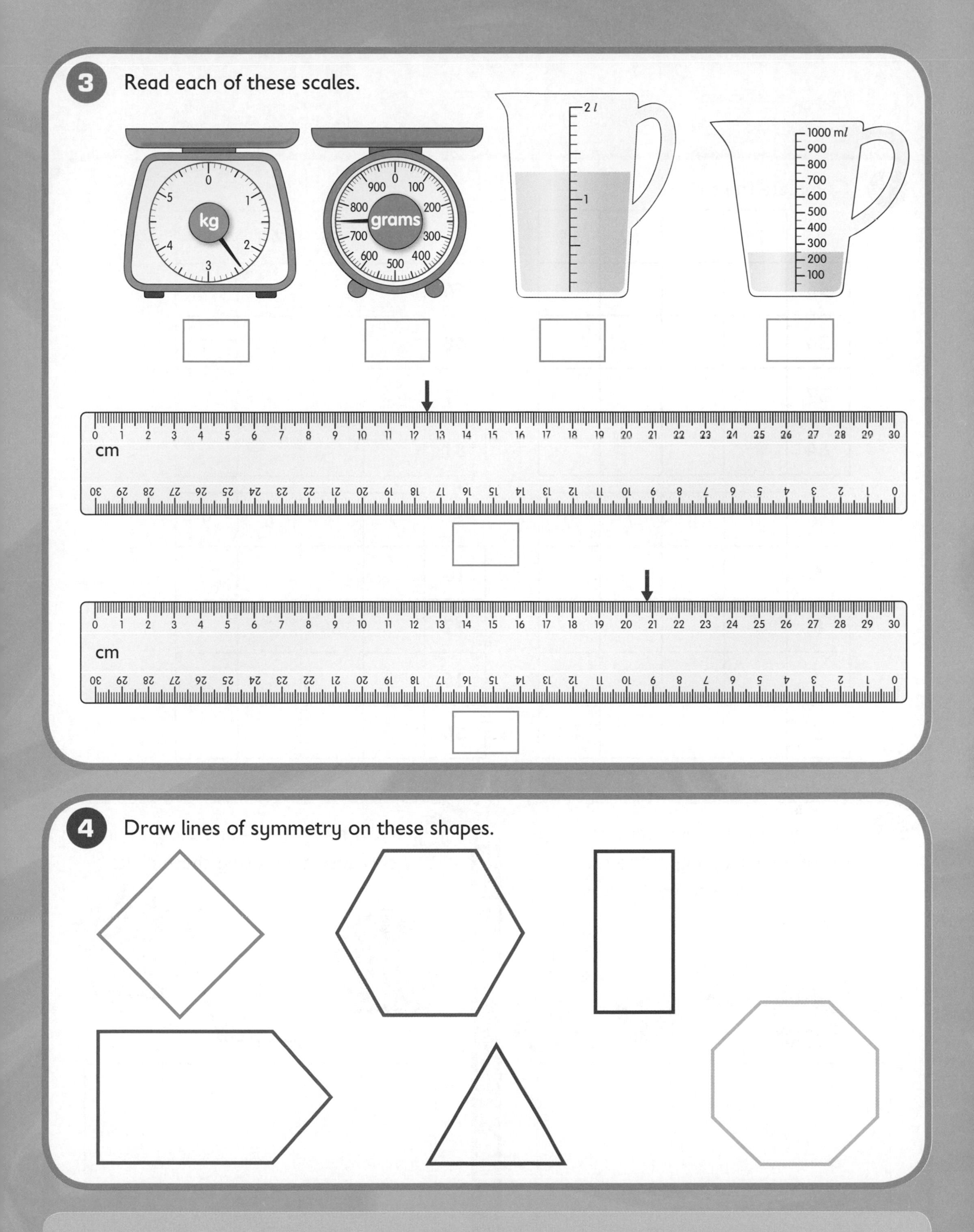

Your child needs to be able to name, and know the properties of, a range of 2-D and 3-D shapes. They also need to be able to estimate, compare and measure lengths, weights and capacities, and interpret the unnumbered divisions on a scale.

Problems and puzzles 1

1 Complete the tables.

+	28	35	12	19
56				
30			42	
47				
64				

−	13	56	44	27
74				
68		12		
97				
81				

×	4	8	7	9
6				
8				
3				
5				45

÷	2	3	6	9
18				
72		24		
36				
54				

2 How many different ways can you make 46, using any of the four operations and some or all of these numbers?

4 3 12 2 8 6

3 How many 1-, 2- and 3-digit numbers can you make using these digits?

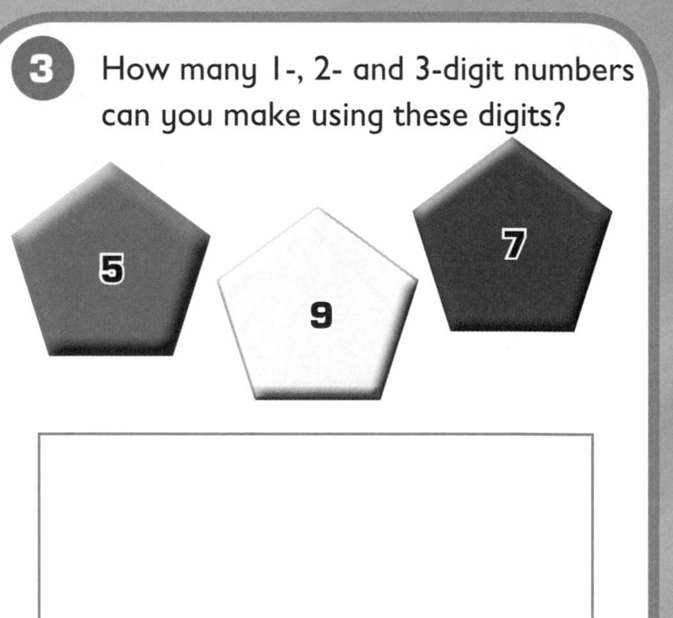

4 Answer these.

A bucket holds 48 litres of water. How many 6 litre jugs can you fill? [ ]

There are 48 chocolates in a box. One-third of the chocolates are dark chocolate and the rest are milk chocolate. How many milk chocolates are there in the box? [ ]

Louise has 6 metres of lace. She cuts it into 8 equal pieces. If Louise uses 3 pieces for her dress, how much lace is left? [ ]

Samson is 114 cm tall. His older brother is 47 cm taller. How tall is Samson's brother? [ ]

There are 72 people on a bus. 33 people are sitting upstairs, 28 people are sitting downstairs and the rest are standing. How many people are standing? [ ]

Thomas spends £1.32 on fruit and £2.85 on a pen. How much does he have left from £5? [ ]

5 Write the digits 3 to 9 on the grid so that each row and column of three digits totals the numbers in the stars.

6 The number in each star is the difference between the two numbers in the circles either side of it. Write the missing numbers in the circles and stars.

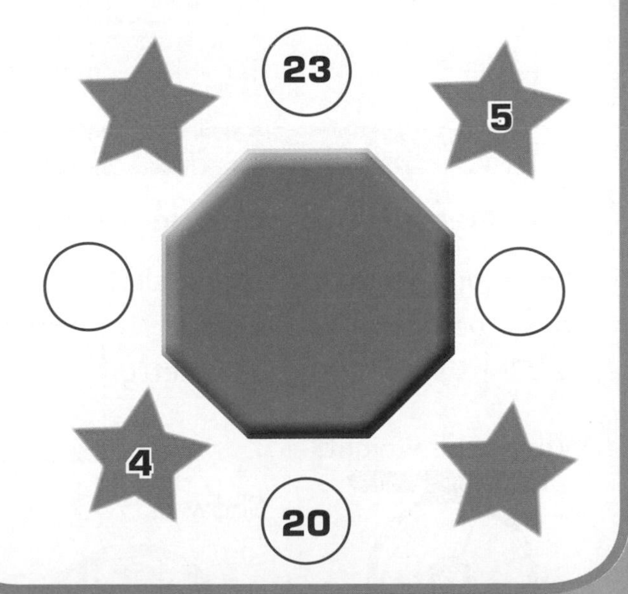

Your child needs to be able to use and apply their mathematical knowledge to solve problems and puzzles in the real world. Wherever possible, ask your child word problems similar to those in **4**.

Quick check 1

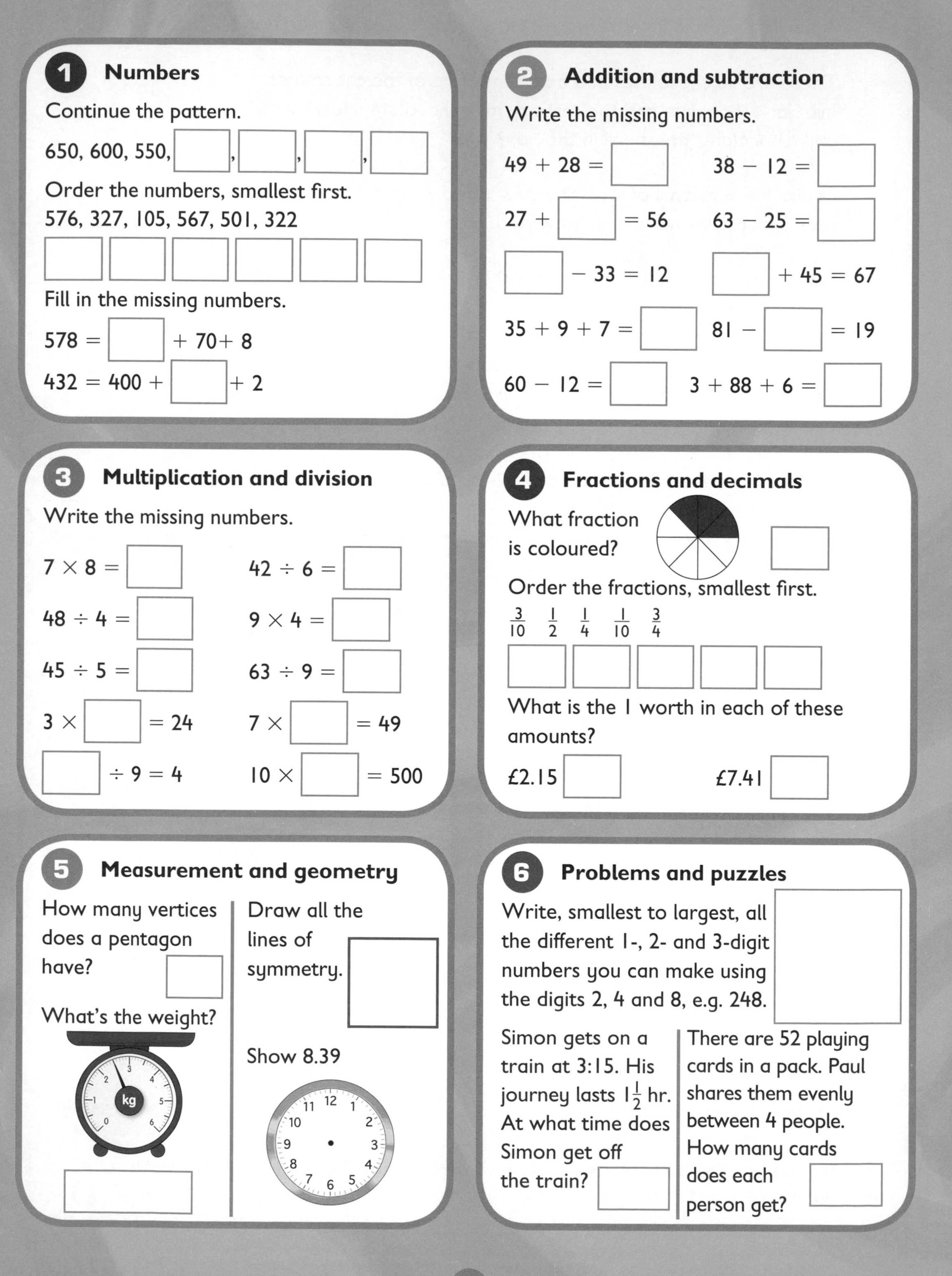

1 Numbers

Continue the pattern.

650, 600, 550, ☐, ☐, ☐, ☐

Order the numbers, smallest first.
576, 327, 105, 567, 501, 322

☐ ☐ ☐ ☐ ☐ ☐

Fill in the missing numbers.

578 = ☐ + 70 + 8

432 = 400 + ☐ + 2

2 Addition and subtraction

Write the missing numbers.

49 + 28 = ☐ 38 − 12 = ☐

27 + ☐ = 56 63 − 25 = ☐

☐ − 33 = 12 ☐ + 45 = 67

35 + 9 + 7 = ☐ 81 − ☐ = 19

60 − 12 = ☐ 3 + 88 + 6 = ☐

3 Multiplication and division

Write the missing numbers.

7 × 8 = ☐ 42 ÷ 6 = ☐

48 ÷ 4 = ☐ 9 × 4 = ☐

45 ÷ 5 = ☐ 63 ÷ 9 = ☐

3 × ☐ = 24 7 × ☐ = 49

☐ ÷ 9 = 4 10 × ☐ = 500

4 Fractions and decimals

What fraction is coloured? ☐

Order the fractions, smallest first.

$\frac{3}{10}$ $\frac{1}{2}$ $\frac{1}{4}$ $\frac{1}{10}$ $\frac{3}{4}$

☐ ☐ ☐ ☐ ☐

What is the 1 worth in each of these amounts?

£2.15 ☐ £7.41 ☐

5 Measurement and geometry

How many vertices does a pentagon have? ☐

What's the weight?

☐

Draw all the lines of symmetry. ☐

Show 8.39

6 Problems and puzzles

Write, smallest to largest, all the different 1-, 2- and 3-digit numbers you can make using the digits 2, 4 and 8, e.g. 248. ☐

Simon gets on a train at 3:15. His journey lasts $1\frac{1}{2}$ hr. At what time does Simon get off the train? ☐

There are 52 playing cards in a pack. Paul shares them evenly between 4 people. How many cards does each person get? ☐

Progress test 1

Time how long it takes you to answer the following questions.

1. $38 + 7 + 4 =$ ☐
2. Order the numbers, smallest first.
   546, 745, 576, 754, 364
   ☐ ☐ ☐ ☐ ☐
3. $6 \times 3 =$ ☐
4. What is the value of the 8 in 582?
   ☐
5. $78 - 47 =$ ☐
6. Continue the pattern.
   387, 391, 395, ☐, ☐, ☐
7. $96 \div 8 =$ ☐
8. $52 \times 100 =$ ☐
9. What fraction is shaded?
   ☐
10. $65 + 80 =$ ☐
11. What is the time?
    ☐
12. $77 - 39 =$ ☐
13. $6800 \div 100 =$ ☐
14. Circle the largest fraction.
    $\frac{3}{4}$ $\frac{1}{10}$ $\frac{1}{2}$
15. Show 350 ml

    500 ml
    400
    300
    200
    100
16. $450 \times 10 =$ ☐
17. £1.43 + £5.54 = ☐
18. How many lines of symmetry does a regular octagon have?
    ☐
19. $279 = 200 +$ ☐ $+ 9$
20. $67 + 28 =$ ☐

Score	Time

Numbers 2

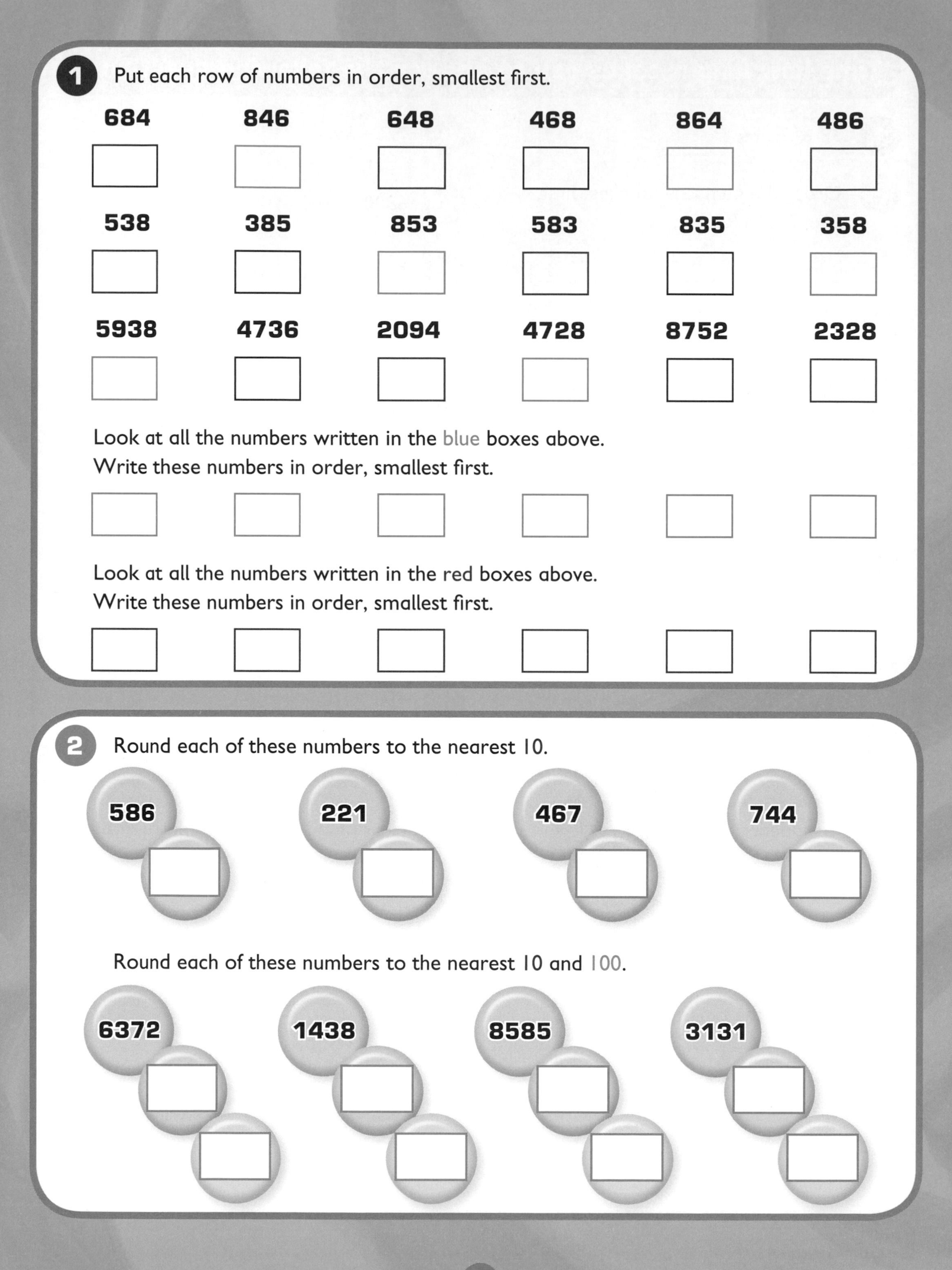

1 Put each row of numbers in order, smallest first.

684 846 648 468 864 486

538 385 853 583 835 358

5938 4736 2094 4728 8752 2328

Look at all the numbers written in the blue boxes above.
Write these numbers in order, smallest first.

Look at all the numbers written in the red boxes above.
Write these numbers in order, smallest first.

2 Round each of these numbers to the nearest 10.

586 221 467 744

Round each of these numbers to the nearest 10 and 100.

6372 1438 8585 3131

Game: Round to 10

You need: two 1–6 dice, counters

- Before you start, decide who will have which colour tower.
- Take turns to:
  - roll the dice and use the two numbers to make a two-digit number,

    e.g. [dice showing 6 and 2] could be either

    62 or 26
  - round the number to the nearest multiple of 10, e.g. 62 would be 60.
  - put a counter on that multiple on your tower.
- The winner is the first player to put a counter on each of their multiples of 10.

70	70
60	60
50	50
40	40
30	30
20	20
10	10

3 What is the value of the red digit in each of these numbers?

2038	☐	**8135**	☐
7461	☐	**1098**	☐
5473	☐	**3535**	☐
4837	☐	**6202**	☐

4 Complete each number sentence.

4625 = 4000 + ☐ + 20 + 5

1284 = ☐ + 200 + 80 + 4

3917 = 3000 + ☐ + ☐ + 7

5871 = 5000 + ☐ + ☐ + 1

8259 = ☐ + 200 + ☐ + 9

☐ = 3000 + 800 + 70 + 2

Being able to round numbers to the nearest 10 and 100 will help your child with estimating answers to calculations involving three-digit numbers, e.g. 547 + 321 ≈ 550 + 320 = 870.

Addition and subtraction 2

You can use known addition and subtraction number facts to help work out the answers when adding or subtracting multiples of 10, 100 and 1000.

Example

$7 + 5 = 12$

So: $70 + 50 = 120$

$700 + 500 = 1200$

$7000 + 5000 = 12\,000$

1 Add together pairs of numbers next to each other and write the answer in the box above.

Find the difference between pairs of numbers next to each other and write the answer in the box above.

2 Add the numbers on the two balls together.

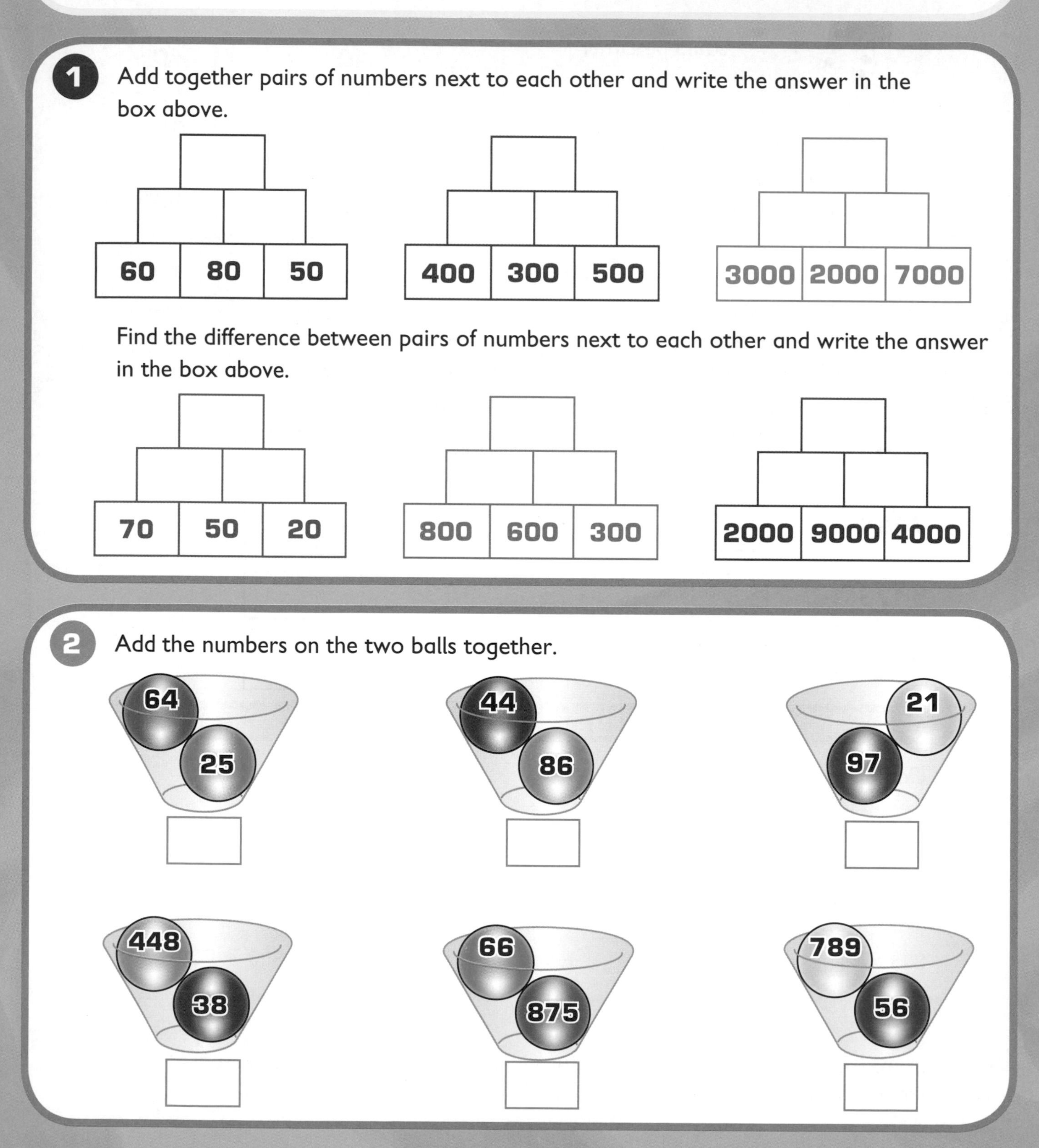

Game: Adding to or subtracting from the stars

You need: pencil and paperclip (for the spinner), counters

490 533 576 619 662

18 61 49 92 77 34

- Take turns to:
  - say a number on a star
  - spin the spinner
  - add or subtract the spinner number to or from the star number
  - say the calculation and put a counter on that answer on the grid.
- If you can't find that answer on the grid, miss that turn.
- The winner is the first player to complete a line of three counters. A line can go horizontally or vertically.

472	601	723	628	515	637
668	524	610	551	696	542
613	499	548	539	653	429
527	398	711	644	441	585
413	570	456	508	770	484
739	680	594	582	625	567

3 Find the difference between the two balls.

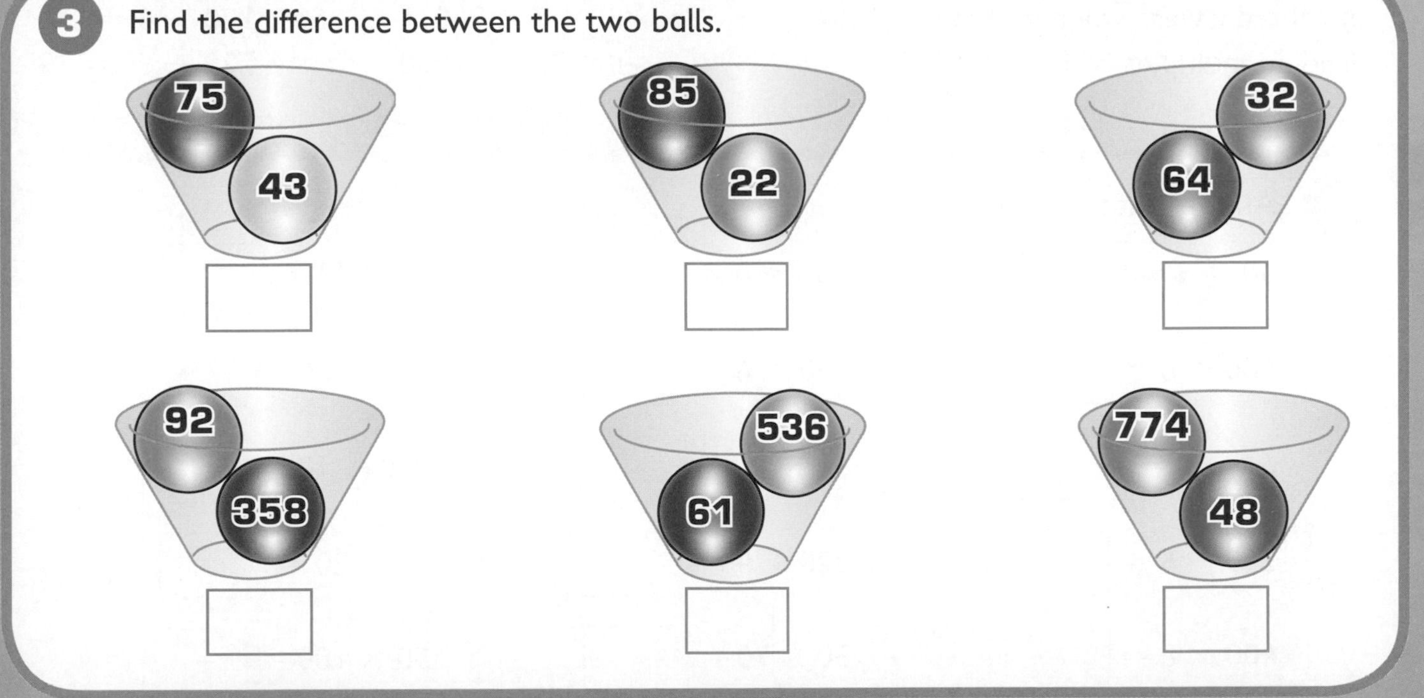

When adding and subtracting pairs of two-digit numbers such as 56 + 39 and 84 − 37, your child may not at first be able to do this entirely in their head. If so, encourage them to make jottings to help them remember what they are doing.

Multiplication and division 2

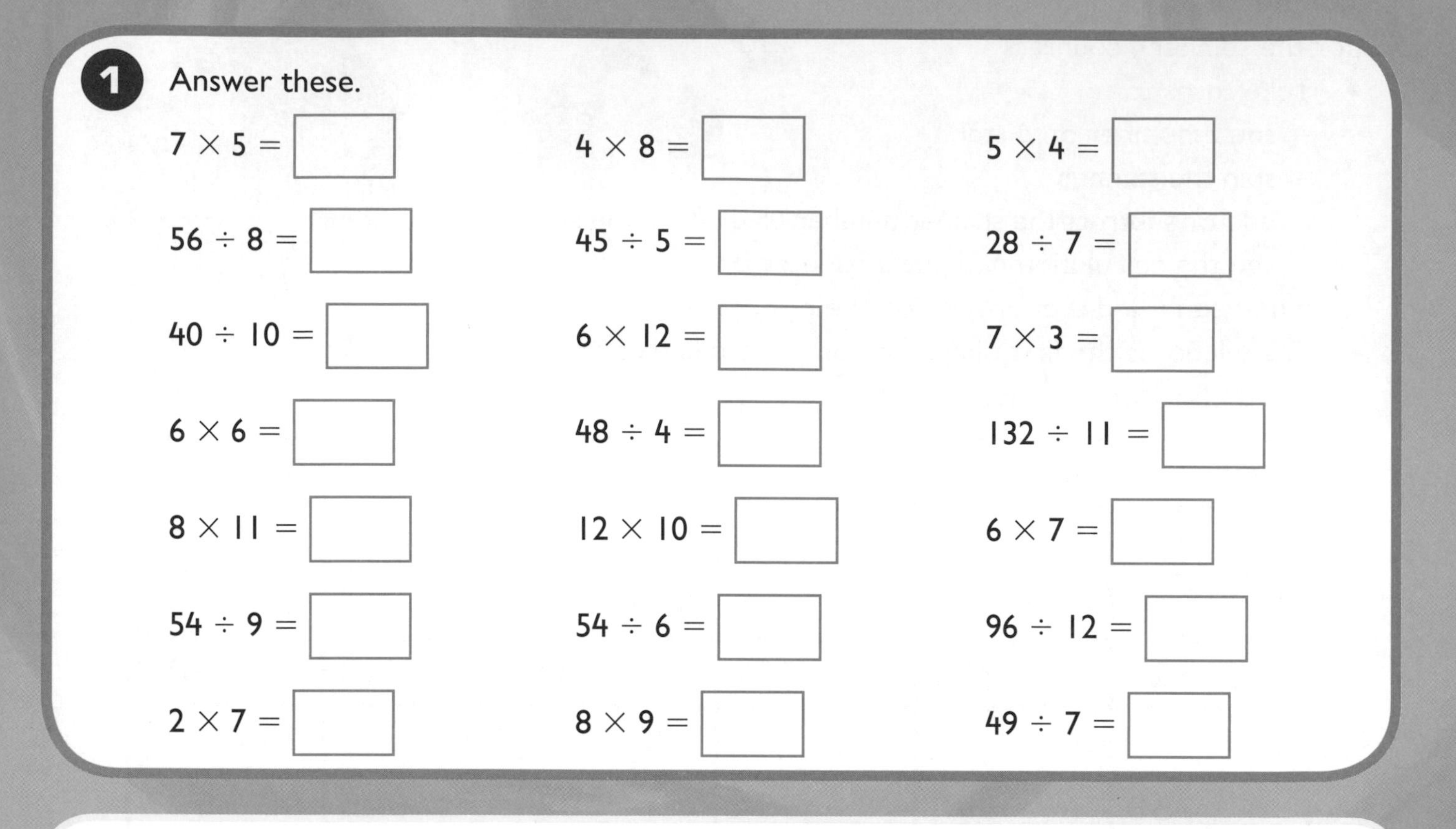

1 Answer these.

$7 \times 5 = \square$	$4 \times 8 = \square$	$5 \times 4 = \square$
$56 \div 8 = \square$	$45 \div 5 = \square$	$28 \div 7 = \square$
$40 \div 10 = \square$	$6 \times 12 = \square$	$7 \times 3 = \square$
$6 \times 6 = \square$	$48 \div 4 = \square$	$132 \div 11 = \square$
$8 \times 11 = \square$	$12 \times 10 = \square$	$6 \times 7 = \square$
$54 \div 9 = \square$	$54 \div 6 = \square$	$96 \div 12 = \square$
$2 \times 7 = \square$	$8 \times 9 = \square$	$49 \div 7 = \square$

You can use known multiplication and division number facts to help work out the answers when multiplying and dividing multiples of 10 and 100.

Example

$4 \times 7 = 28$

So: $40 \times 7 = 280$

$4 \times 700 = 2800$

Example

$72 \div 6 = 12$

So: $720 \div 6 = 120$

$720 \div 60 = 12$

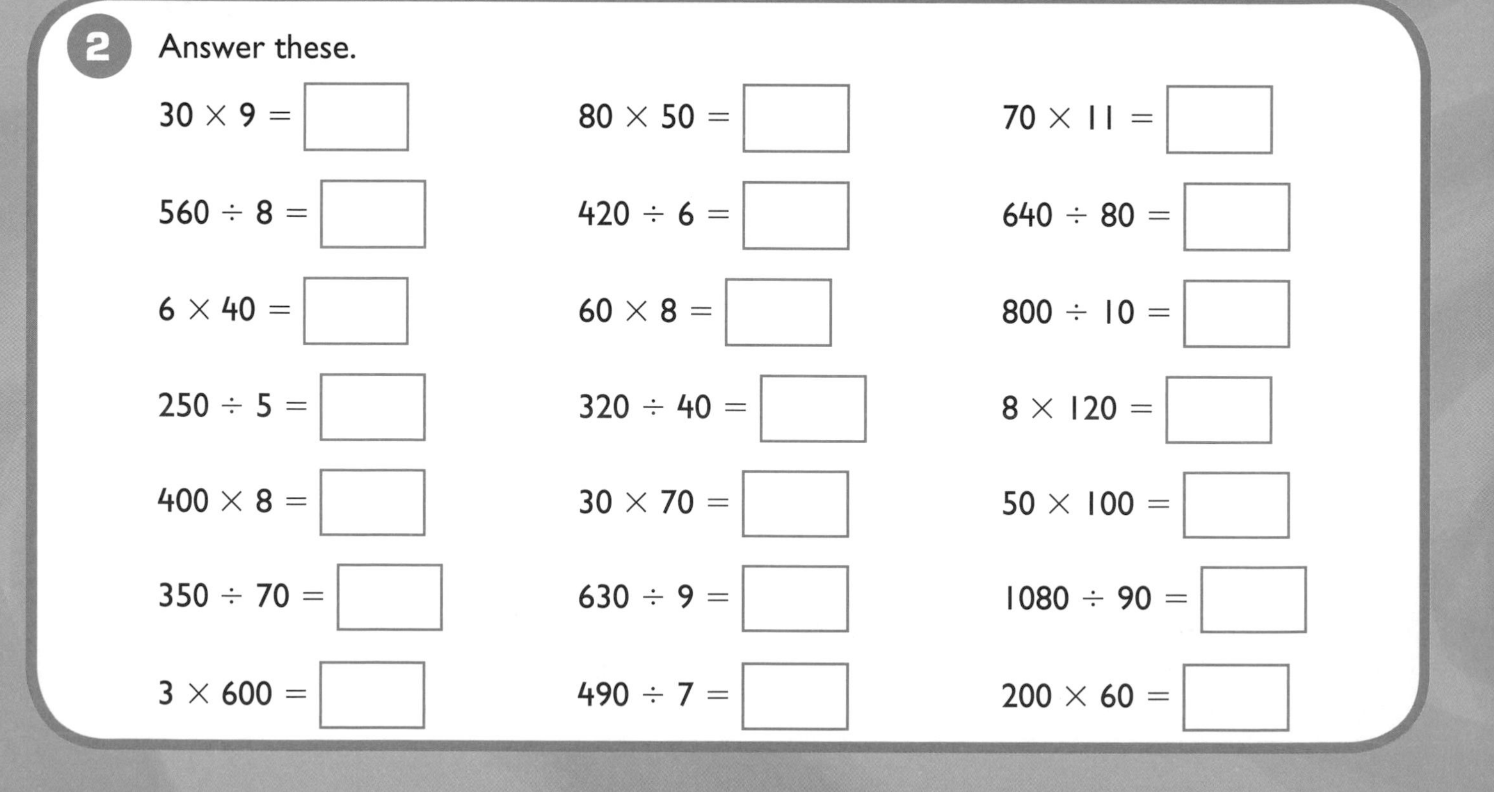

2 Answer these.

$30 \times 9 = \square$	$80 \times 50 = \square$	$70 \times 11 = \square$
$560 \div 8 = \square$	$420 \div 6 = \square$	$640 \div 80 = \square$
$6 \times 40 = \square$	$60 \times 8 = \square$	$800 \div 10 = \square$
$250 \div 5 = \square$	$320 \div 40 = \square$	$8 \times 120 = \square$
$400 \times 8 = \square$	$30 \times 70 = \square$	$50 \times 100 = \square$
$350 \div 70 = \square$	$630 \div 9 = \square$	$1080 \div 90 = \square$
$3 \times 600 = \square$	$490 \div 7 = \square$	$200 \times 60 = \square$

Game: The multiples game

You need: pencil and paperclip (for the spinner),
20 counters: 10 of one colour, 10 of another colour

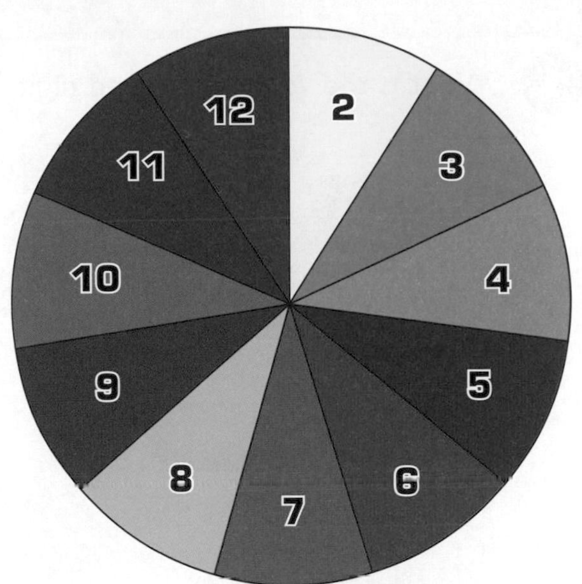

- Before you start choose who will have which colour counters.
- Take turns to place one of your counters on a number on the grid, making sure not to cover up the number.
- Keep going until you have each placed all your 10 counters on the grid.

Rule: Only one counter can go on each number.

- Now take turns to:
  - spin the spinner
  - take a counter off the grid showing a multiple of that number, but only if the counter is in your colour.
- The winner is the first player to collect all their counters.

66	45	10	28	33	63
6	8	36	20	48	15
35	12	32	22	18	56
40	54	60	14	72	88
24	27	42	30	21	16

3 Double each of these numbers.

54 → ☐ 76 → ☐ 60 → ☐ 89 → ☐ 300 → ☐

Halve each of these numbers.

82 → ☐ 94 → ☐ 78 → ☐ 700 → ☐ 56 → ☐

Using knowledge of their times tables facts, as well as being able to multiply multiples of 10 and 100, will help your child when multiplying larger numbers such as 62 × 8 and 268 × 7.

Fractions and decimals 2

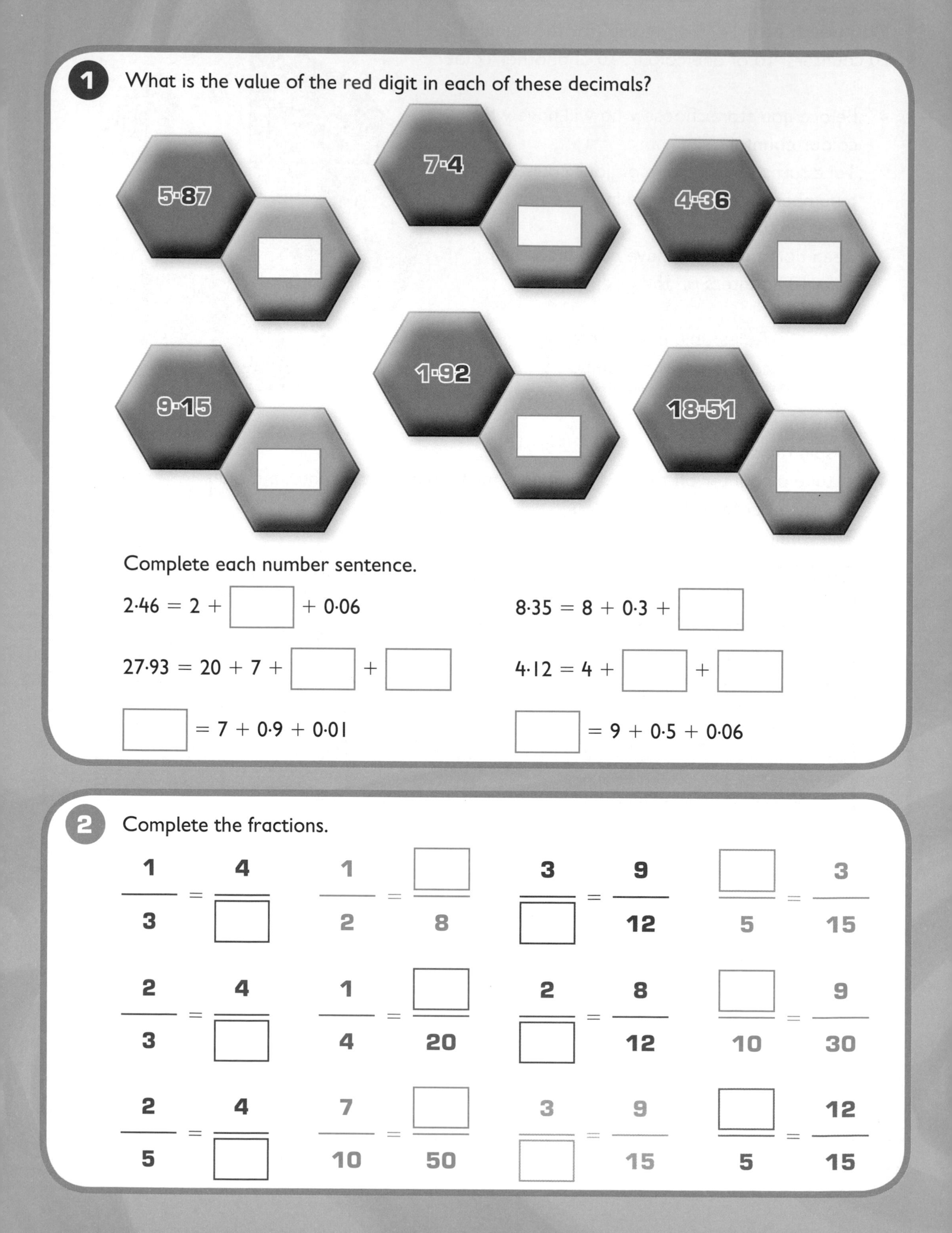

Game: Finding tenths

You need: two 1–6 dice, counters

- Take turns to:
  - roll both dice, e.g. 2 and 5 (if you roll a double, roll the dice again)
  - place one of your counters on the grid on a decimal which is between the two dice numbers, e.g. 4·7.
- The winner is the first player to complete a line of four counters. A line can go horizontally, vertically or diagonally.

2·6	6·2	4·8	5·4	1·2	3·9
6·4	4·1	3·3	1·6	2·9	5·5
4·7	2·5	1·1	3·2	5·8	6·7
1·7	5·1	6·9	2·1	3·6	4·4
3·4	1·8	5·8	6·6	4·2	2·3
5·6	3·1	2·7	4·3	6·5	1·4

3 Change these fractions to decimals.

$\frac{1}{2}$ = ☐ $\frac{1}{10}$ = ☐ $\frac{1}{4}$ = ☐ $\frac{6}{10}$ = ☐

$\frac{3}{4}$ = ☐ $\frac{3}{10}$ = ☐ $\frac{8}{10}$ = ☐ $\frac{4}{10}$ = ☐

$2\frac{2}{10}$ = ☐ $3\frac{7}{10}$ = ☐ $1\frac{9}{10}$ = ☐ $5\frac{3}{4}$ = ☐

Your child needs to be able to recognise equivalent fractions as well as recognise and write decimal equivalents for ¼, ½, ¾ and tenths. They also need to have a secure understanding of place value for decimal numbers up to two decimal places, like the red numbers in **1**.

Measurement and geometry 2

Use a red pencil to show a horizontal line in each of the following pictures. Use a blue pencil to show a vertical line in each picture.

2 Use a green pencil to show a pair of parallel lines in each of the following pictures. Use an purple pencil to show a pair of perpendicular lines in each picture.

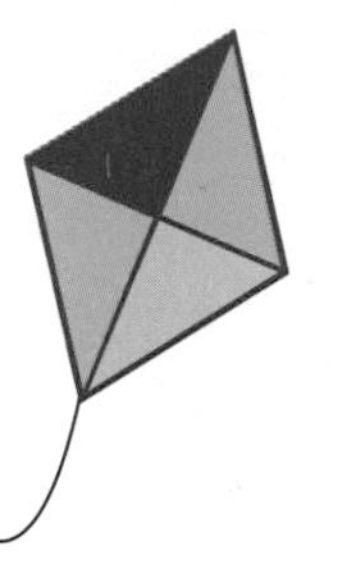

3 Name each angle and then order them, smallest to largest.

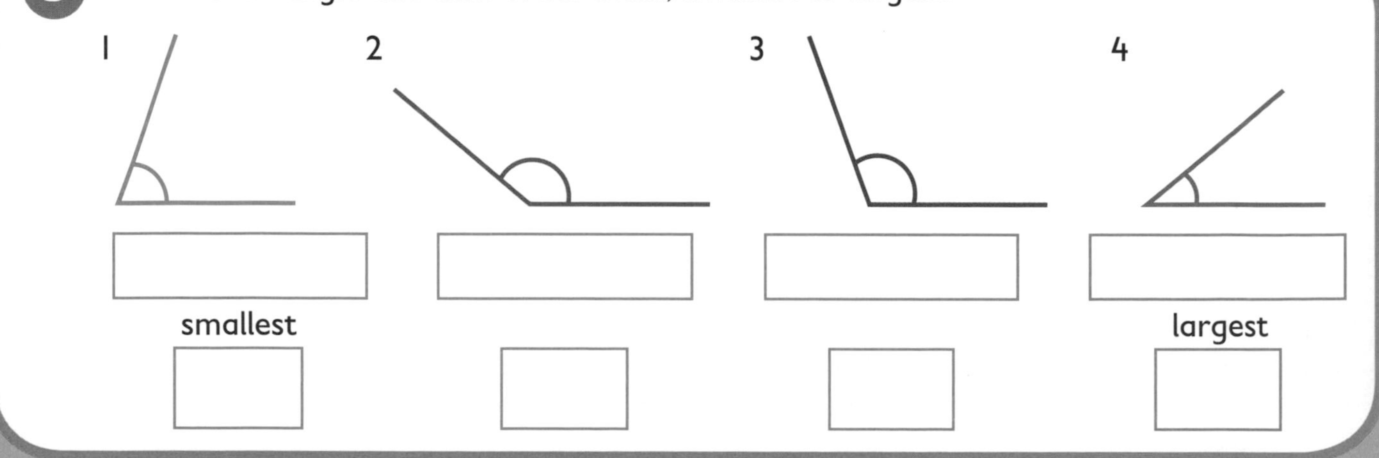

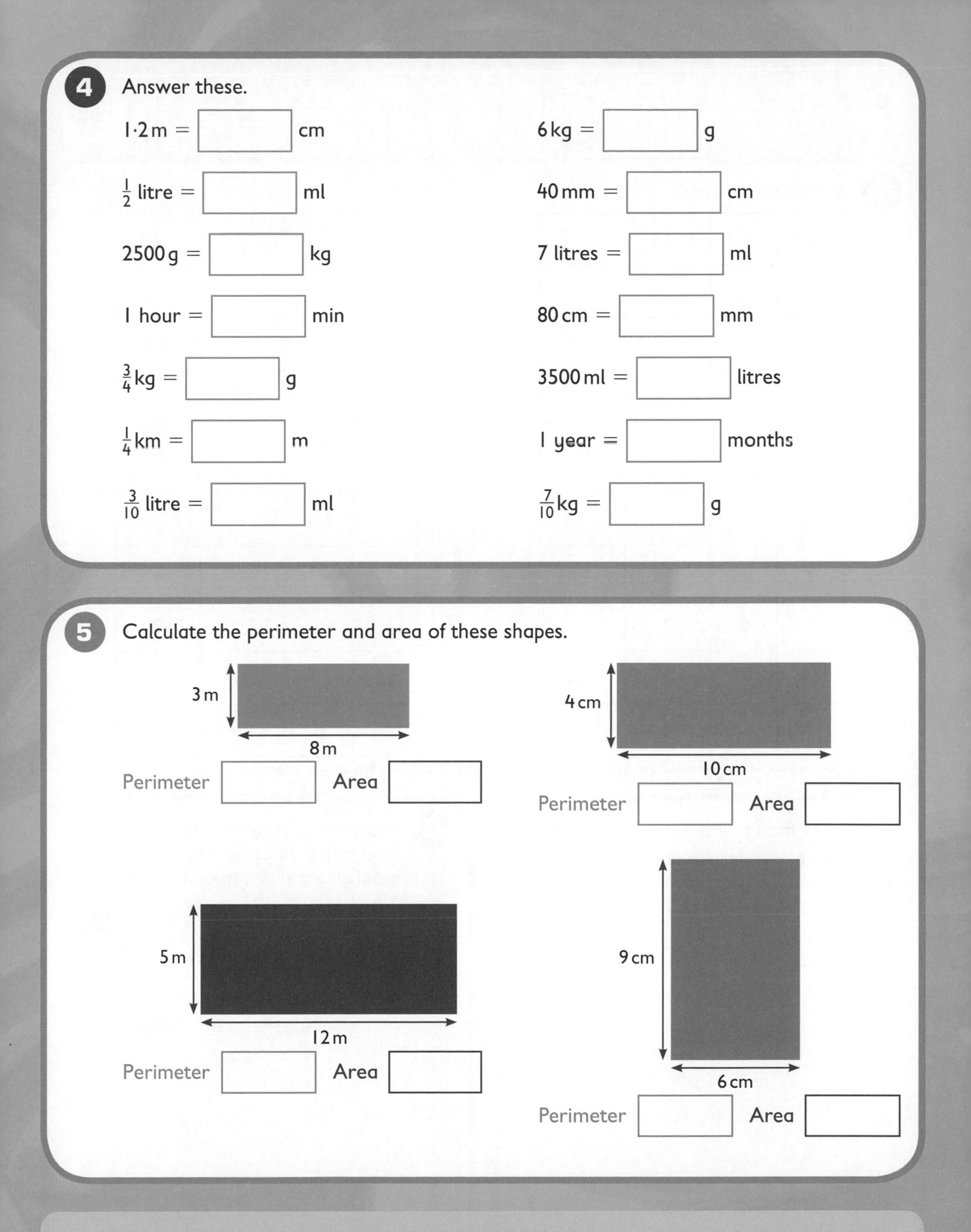

4 Answer these.

$1{\cdot}2$ m = ☐ cm

$\frac{1}{2}$ litre = ☐ ml

2500 g = ☐ kg

1 hour = ☐ min

$\frac{3}{4}$ kg = ☐ g

$\frac{1}{4}$ km = ☐ m

$\frac{3}{10}$ litre = ☐ ml

6 kg = ☐ g

40 mm = ☐ cm

7 litres = ☐ ml

80 cm = ☐ mm

3500 ml = ☐ litres

1 year = ☐ months

$\frac{7}{10}$ kg = ☐ g

5 Calculate the perimeter and area of these shapes.

3 m
8 m

Perimeter ☐ Area ☐

4 cm
10 cm

Perimeter ☐ Area ☐

5 m
12 m

Perimeter ☐ Area ☐

9 cm
6 cm

Perimeter ☐ Area ☐

Your child needs to be able to use standard units for length, mass, capacity and time and to be able to convert between different units of measurement. They should also be able to work out the perimeter and area of simple rectilinear shapes.

Problems and puzzles 2

Complete the grids.

◇	○	◇+○	◇−○
15	12		
17	11		
25			4
	25		11

◇	○	◇+○	◇×○
6	3		
8			56
4		13	
	12		60

◇	○	◇−○	◇×○
9			18
	8	2	
11			121
	4	4	

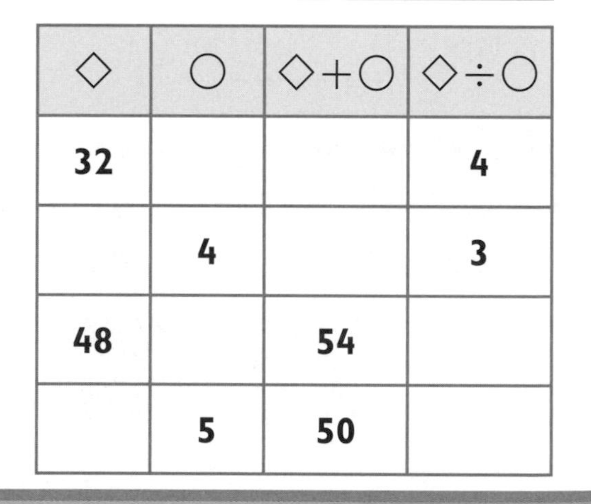

◇	○	◇+○	◇÷○
32			4
	4		3
48		54	
	5	50	

2 How many different ways can you make $5\frac{1}{2}$? Here is one example.

$$8 - 2\frac{1}{2} = 5\frac{1}{2}$$

3 How many different ways can you make 25 using any of the four operations and some or all of these numbers?

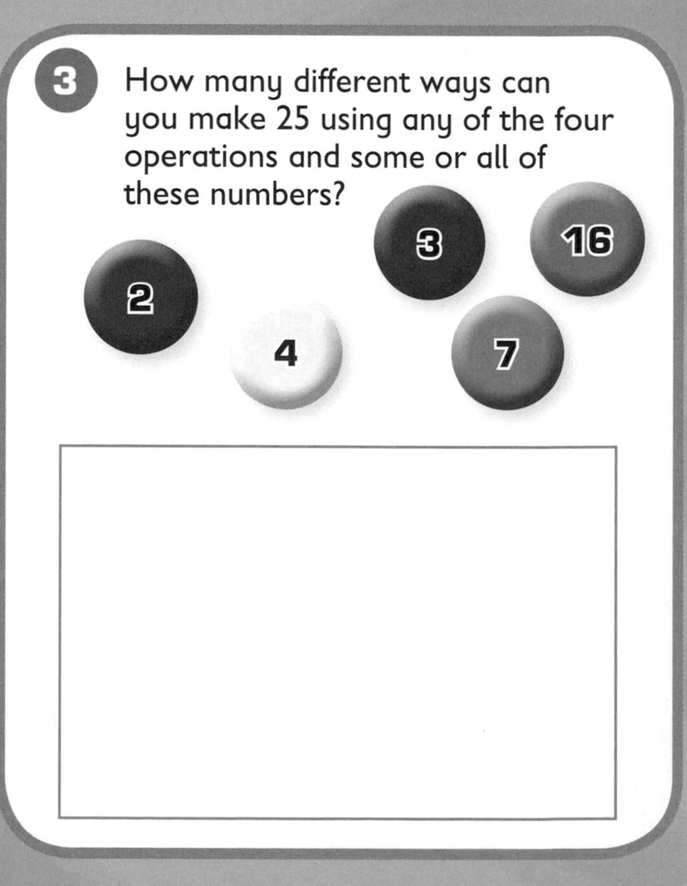

4 Answer these.

Fiona starts her homework at 4:05 pm and finishes at 5:25 pm. How long does Fiona spend on her homework? ☐

There were 84 people at a party. Of these $\frac{3}{4}$ were adults and the rest were children. How many children were at the party? ☐

Henry the chef is preparing for a party and needs to use 100 eggs. How many boxes of 12 eggs does he need to buy? ☐

A supermarket has specials on yogurt and orange juice: buy 2 orange juice for £3 and 4 yogurts for £1.50. Leroy buys 4 orange juice and 8 yogurts. How much change does he get back from £20? ☐

On Friday, Lee drove 120 km from his home to his parents' home. On Saturday, he drove a further 45 km to visit his aunt. On Sunday, he drove back home from his aunt's home via his parents' home. Altogether how many kilometres did he drive at the weekend? ☐

5 Using each of the digits 1 to 9 only once, complete these 4 times tables facts.

4 1 2 3 5 7 8 6 9

☐ × ☐ = 56

☐ × ☐ = 24

☐ × ☐ = 27

☐☐ × ☐ = 60

6 Write each of the digits 1 to 5 once only in each row and column. Use the < and > signs to help you.

☐		☐	>	☐		☐	>	☐
☐	>	☐		☐		4		☐
☐		☐		3		☐		☐
☐	<	☐		☐		☐		☐
						∧		∧
3		☐		☐	>	☐		☐

Find times at home, when out shopping or visiting other places to ask your child a word problem like those in **4**. Try not to just focus on the answer to the problem, but talk to your child about the way they work things out.

Quick check 2

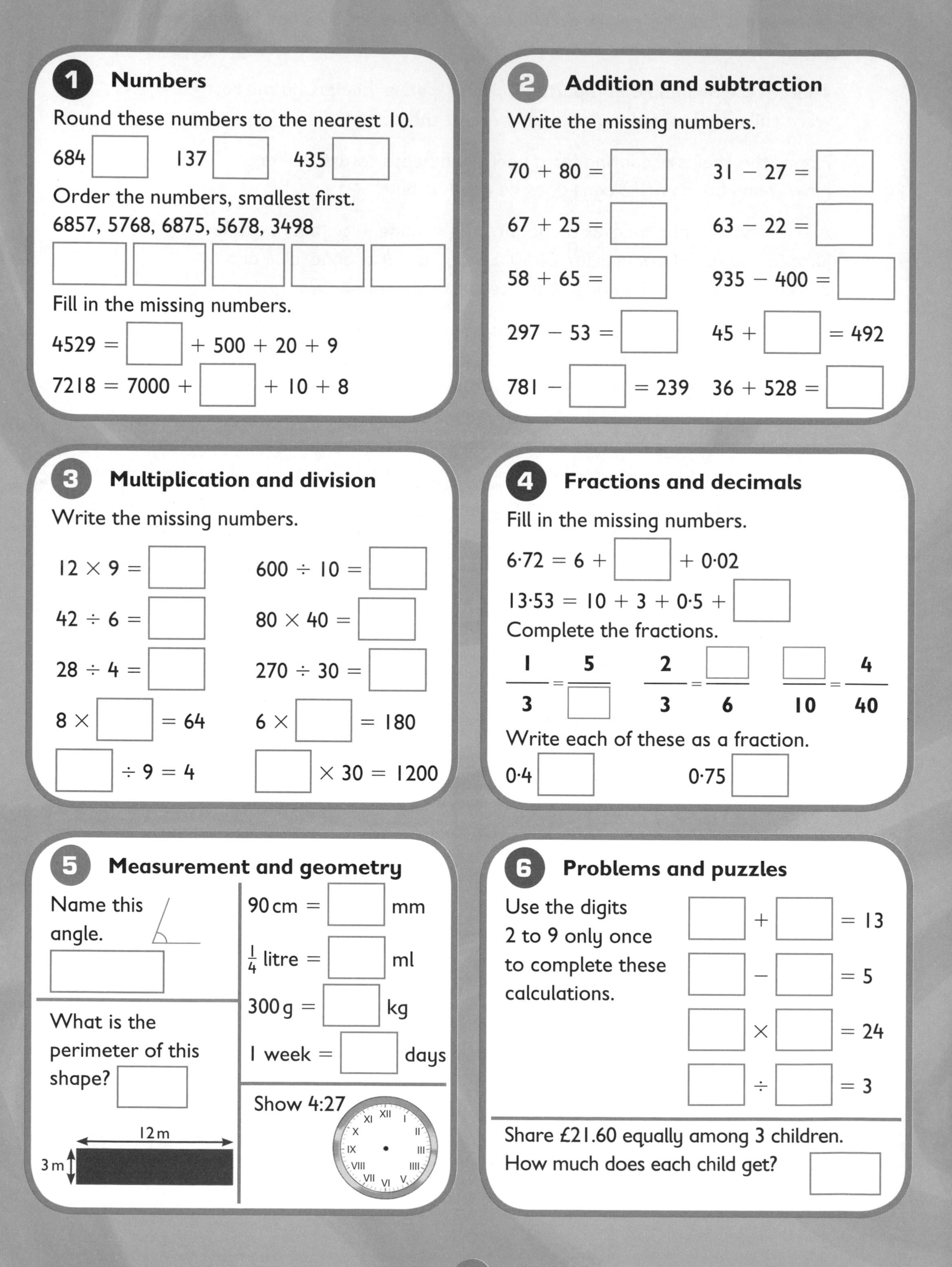

1 Numbers

Round these numbers to the nearest 10.

684 ☐ 137 ☐ 435 ☐

Order the numbers, smallest first.

6857, 5768, 6875, 5678, 3498

☐ ☐ ☐ ☐ ☐

Fill in the missing numbers.

4529 = ☐ + 500 + 20 + 9

7218 = 7000 + ☐ + 10 + 8

2 Addition and subtraction

Write the missing numbers.

70 + 80 = ☐

31 − 27 = ☐

67 + 25 = ☐

63 − 22 = ☐

58 + 65 = ☐

935 − 400 = ☐

297 − 53 = ☐

45 + ☐ = 492

781 − ☐ = 239

36 + 528 = ☐

3 Multiplication and division

Write the missing numbers.

12 × 9 = ☐

600 ÷ 10 = ☐

42 ÷ 6 = ☐

80 × 40 = ☐

28 ÷ 4 = ☐

270 ÷ 30 = ☐

8 × ☐ = 64

6 × ☐ = 180

☐ ÷ 9 = 4

☐ × 30 = 1200

4 Fractions and decimals

Fill in the missing numbers.

6·72 = 6 + ☐ + 0·02

13·53 = 10 + 3 + 0·5 + ☐

Complete the fractions.

$\frac{1}{3} = \frac{5}{\square}$ $\frac{2}{3} = \frac{\square}{6}$ $\frac{\square}{10} = \frac{4}{40}$

Write each of these as a fraction.

0·4 ☐ 0·75 ☐

5 Measurement and geometry

Name this angle. ☐

What is the perimeter of this shape? ☐

90 cm = ☐ mm

$\frac{1}{4}$ litre = ☐ ml

300 g = ☐ kg

1 week = ☐ days

Show 4:27

6 Problems and puzzles

Use the digits 2 to 9 only once to complete these calculations.

☐ + ☐ = 13

☐ − ☐ = 5

☐ × ☐ = 24

☐ ÷ ☐ = 3

Share £21.60 equally among 3 children.

How much does each child get? ☐

Progress test 2

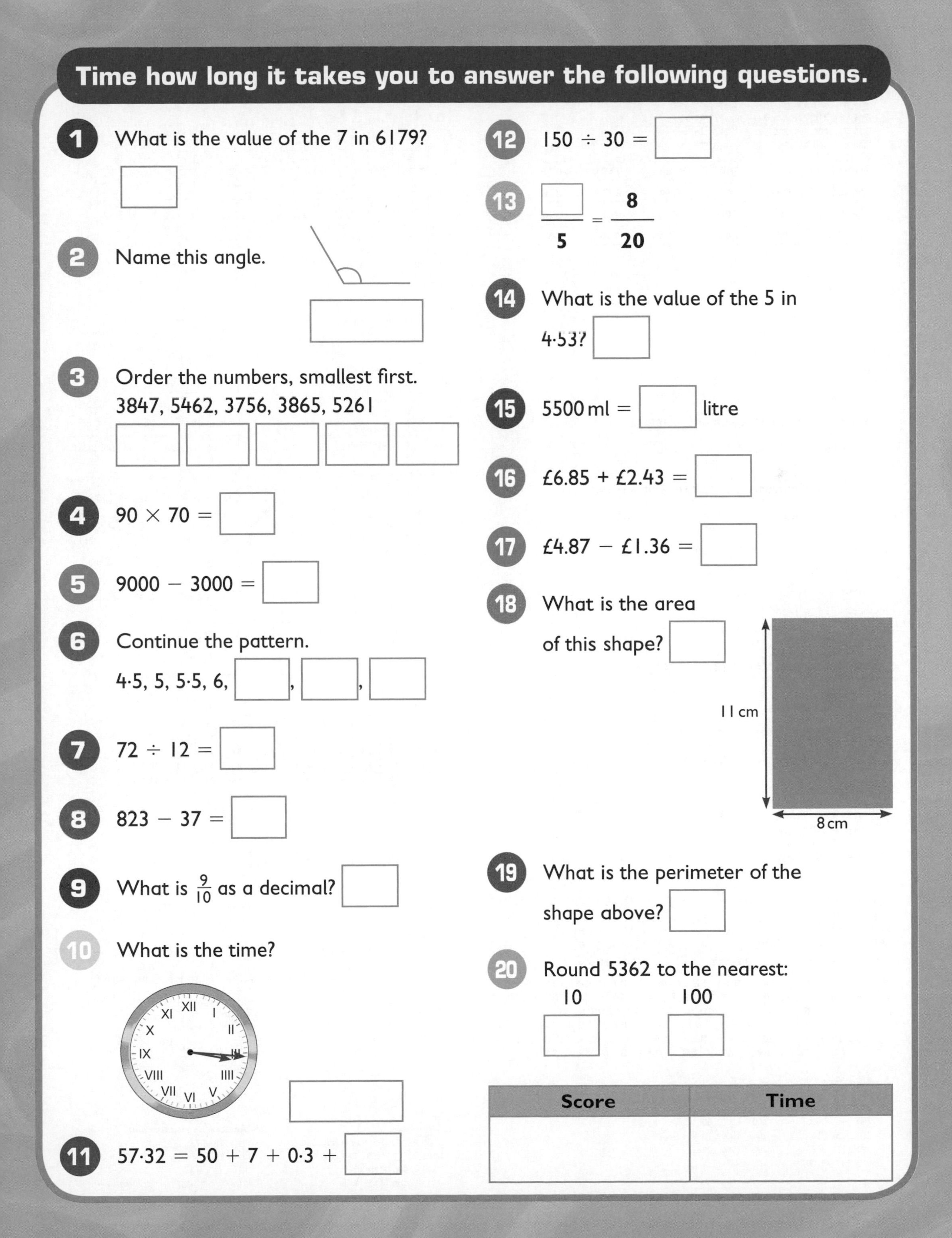

Time how long it takes you to answer the following questions.

1. What is the value of the 7 in 6179? ☐
2. Name this angle. ☐
3. Order the numbers, smallest first.
   3847, 5462, 3756, 3865, 5261
   ☐ ☐ ☐ ☐ ☐
4. $90 \times 70 =$ ☐
5. $9000 - 3000 =$ ☐
6. Continue the pattern.
   4·5, 5, 5·5, 6, ☐, ☐, ☐
7. $72 \div 12 =$ ☐
8. $823 - 37 =$ ☐
9. What is $\frac{9}{10}$ as a decimal? ☐
10. What is the time? ☐

11. $57{\cdot}32 = 50 + 7 + 0{\cdot}3 +$ ☐
12. $150 \div 30 =$ ☐
13. $\frac{☐}{5} = \frac{8}{20}$
14. What is the value of the 5 in 4·53? ☐
15. 5500 ml = ☐ litre
16. £6.85 + £2.43 = ☐
17. £4.87 − £1.36 = ☐
18. What is the area of this shape? ☐

19. What is the perimeter of the shape above? ☐
20. Round 5362 to the nearest:
    10 ☐ 100 ☐

Score	Time

Answers

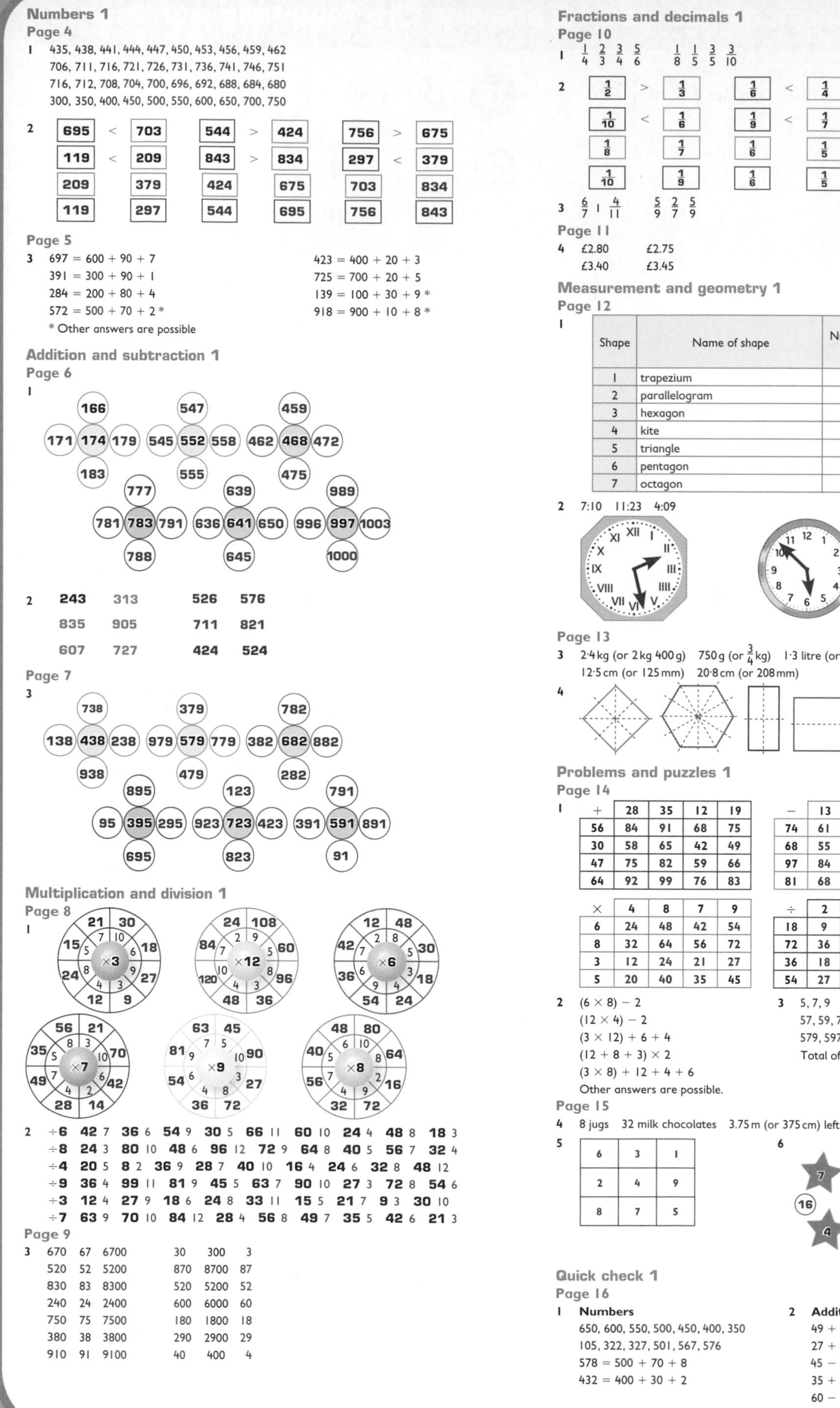

Numbers 1

Page 4

1 435, 438, 441, 444, 447, 450, 453, 456, 459, 462
706, 711, 716, 721, 726, 731, 736, 741, 746, 751
716, 712, 708, 704, 700, 696, 692, 688, 684, 680
300, 350, 400, 450, 500, 550, 600, 650, 700, 750

2 695 < 703 544 > 424 756 > 675
119 < 209 843 > 834 297 < 379
209 379 424 675 703 834
119 297 544 695 756 843

Page 5

3 697 = 600 + 90 + 7
391 = 300 + 90 + 1
284 = 200 + 80 + 4
572 = 500 + 70 + 2 *
423 = 400 + 20 + 3
725 = 700 + 20 + 5
139 = 100 + 30 + 9 *
918 = 900 + 10 + 8 *
* Other answers are possible

Addition and subtraction 1

Page 6

1 166, 171, 174, 179, 183 547, 545, 552, 558, 555 459, 462, 468, 472, 475
777, 781, 783, 791, 788 639, 636, 641, 650, 645 989, 996, 997, 1003, 1000

2 243 313 526 576
835 905 711 821
607 727 424 524

Page 7

3 738, 138, 438, 238, 938 379, 979, 579, 779, 479 782, 382, 682, 882, 282
895, 95, 395, 295, 695 123, 923, 723, 423, 823 791, 391, 591, 891, 91

Multiplication and division 1

Page 8

1 ×3: 21 30 18 27 9 12 24 15; 7 10 6 9 3 4 8 5
×12: 24 108 60 96 36 48 120 84; 2 9 5 8 3 4 10 7
×6: 12 48 30 18 24 54 36 42; 2 8 5 3 4 9 6 7
×7: 56 21 70 42 14 28 49 35; 8 3 10 6 2 4 7 5
×9: 63 45 90 27 72 36 54 81; 7 5 10 3 8 4 6 9
×8: 48 80 64 16 72 32 56 40; 6 10 8 2 9 4 7 5

2 ÷6 42 7 36 6 54 9 30 5 66 11 60 10 24 4 48 8 18 3
÷8 24 3 80 10 48 6 96 12 72 9 64 8 40 5 56 7 32 4
÷4 20 5 8 2 36 9 28 7 40 10 16 4 24 6 32 8 48 12
÷9 36 4 99 11 81 9 45 5 63 7 90 10 27 3 72 8 54 6
÷3 12 4 27 9 18 6 24 8 33 11 15 5 21 7 9 3 30 10
÷7 63 9 70 10 84 12 28 4 56 8 49 7 35 5 42 6 21 3

Page 9

3 670 67 6700 30 300 3
520 52 5200 870 8700 87
830 83 8300 520 5200 52
240 24 2400 600 6000 60
750 75 7500 180 1800 18
380 38 3800 290 2900 29
910 91 9100 40 400 4

Fractions and decimals 1

Page 10

1 $\frac{1}{4}$ $\frac{2}{3}$ $\frac{3}{4}$ $\frac{5}{6}$ $\frac{1}{8}$ $\frac{1}{5}$ $\frac{3}{5}$ $\frac{3}{10}$

2 $\frac{1}{2}$ > $\frac{1}{3}$ $\frac{1}{6}$ < $\frac{1}{4}$ $\frac{1}{5}$ > $\frac{1}{8}$
$\frac{1}{10}$ < $\frac{1}{6}$ $\frac{1}{9}$ < $\frac{1}{7}$ $\frac{1}{3}$ > $\frac{1}{5}$
$\frac{1}{8}$ $\frac{1}{7}$ $\frac{1}{6}$ $\frac{1}{5}$ $\frac{1}{4}$ $\frac{1}{3}$
$\frac{1}{10}$ $\frac{1}{9}$ $\frac{1}{6}$ $\frac{1}{5}$ $\frac{1}{3}$ $\frac{1}{2}$

3 $\frac{6}{7}$ 1 $\frac{4}{11}$ $\frac{5}{9}$ $\frac{2}{7}$ $\frac{5}{9}$

Page 11

4 £2.80 £2.75
£3.40 £3.45

Measurement and geometry 1

Page 12

1

Shape	Name of shape	Number of sides	Number of vertices	Number of lines of symmetry
1	trapezium	4	4	0
2	parallelogram	4	4	0
3	hexagon	6	6	6
4	kite	4	4	1
5	triangle	3	3	3
6	pentagon	5	5	5
7	octagon	8	8	8

2 7:10 11:23 4:09

Page 13

3 2·4 kg (or 2 kg 400 g) 750 g (or $\frac{3}{4}$ kg) 1·3 litre (or 1 litre 300 ml) 250 ml (or $\frac{1}{4}$ litre)
12·5 cm (or 125 mm) 20·8 cm (or 208 mm)

4

Problems and puzzles 1

Page 14

1

+	28	35	12	19
56	84	91	68	75
30	58	65	42	49
47	75	82	59	66
64	92	99	76	83

−	13	56	44	27
74	61	18	30	47
68	55	12	24	41
97	84	41	53	70
81	68	25	37	54

×	4	8	7	9
6	24	48	42	54
8	32	64	56	72
3	12	24	21	27
5	20	40	35	45

÷	2	3	6	9
18	9	6	3	2
72	36	24	12	8
36	18	12	6	4
54	27	18	9	6

2 (6 × 8) − 2
(12 × 4) − 2
(3 × 12) + 6 + 4
(12 + 8 + 3) × 2
(3 × 8) + 12 + 4 + 6
Other answers are possible.

3 5, 7, 9
57, 59, 75, 79, 95, 97
579, 597, 759, 795, 957, 975
Total of 15 numbers

Page 15

4 8 jugs 32 milk chocolates 3.75 m (or 375 cm) left 161 cm tall 11 people 83p

5

6	3	1
2	4	9
8	7	5

6

Quick check 1

Page 16

1 **Numbers**
650, 600, 550, 500, 450, 400, 350
105, 322, 327, 501, 567, 576
578 = 500 + 70 + 8
432 = 400 + 30 + 2

2 **Addition and subtraction**
49 + 28 = 77 38 − 12 = 26
27 + 29 = 56 63 − 25 = 38
45 − 33 = 12 22 + 45 = 67
35 + 9 + 7 = 51 81 − 62 = 19
60 − 12 = 48 3 + 88 + 6 = 97